LEARNING
Basic Mathematics

I

New Century Edition

William F. Hunter, Ph.D.
Formerly Clinical Psychologist
Minneapolis, Minnesota

Neysa Chouteau
Subscripts Writing and
Editing Resource Group

Ethel G. Armstrong
Formerly teacher, Whitfield School
St. Louis, Missouri

Phoenix Learning Resources, LLC

PO Box 510 • Honesdale, PA 18431
1-800-228-9345 • Fax: 570-253-3227 • www.phoenixlr.com

Item# 3236   ISBN 978-0-7915-3236-2

Contents

✔ = evaluation   R = review

ISBN 0-7915-3236-4

2 3 4 5 6 06 05 04

Name _____ Date _____

Color.

1 – one

☐ Color 1 ball.

1

one ball

☐ Color 1 cup.

1

one cup

1

Color.

☐ Color 1 shoe.

1

one shoe

☐ Color 1 apple.

1

one apple

Name _____ Date _____

Color.

2 – two

☐ Color 2 balls.

2

two balls

☐ Color 2 cups.

2

two cups

Color.

☐ Color 2 shoes.

2

two shoes

☐ Color 2 apples.

2

two apples

Matching – 1 and 2

☐ Draw a line.

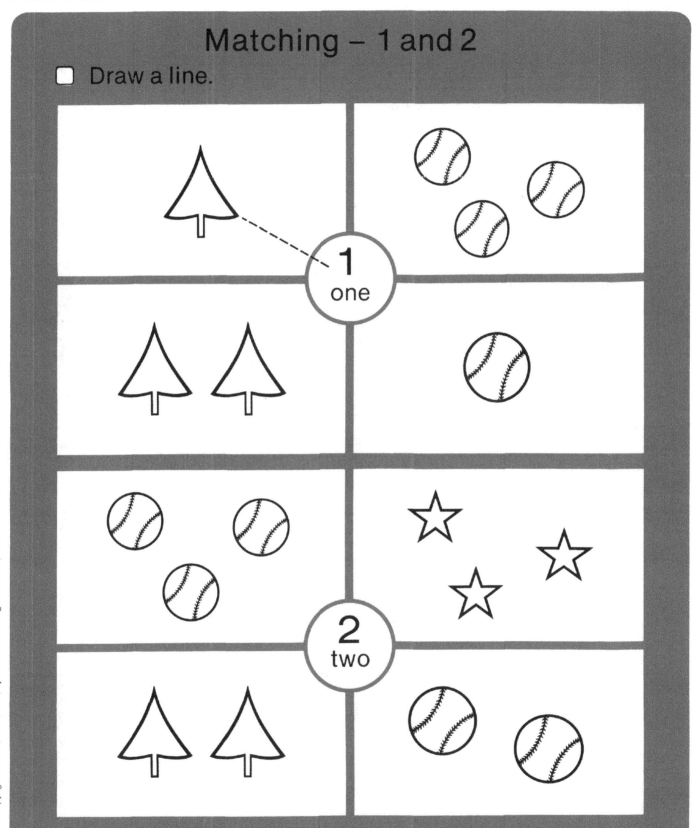

☐ Draw a line.

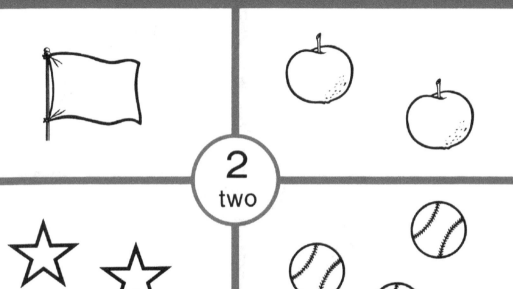

Name _____ Date _____

Matching – 1 and 2

☐ How many?

1 ②

1 2

1 2

1 2

7

✓ Matching – 1 and 2

☐ How many?

1 2

1 2

1 2

1 2

Name _____ Date _____

Color.

3 – three

☐ Color 3 balls.

3

three balls

☐ Color 3 cups.

3

three cups

Color.

☐ Color 3 shoes.

3

three shoes

☐ Color 3 apples.

3

three apples

Matching – 1 and 3

☐ Draw a line.

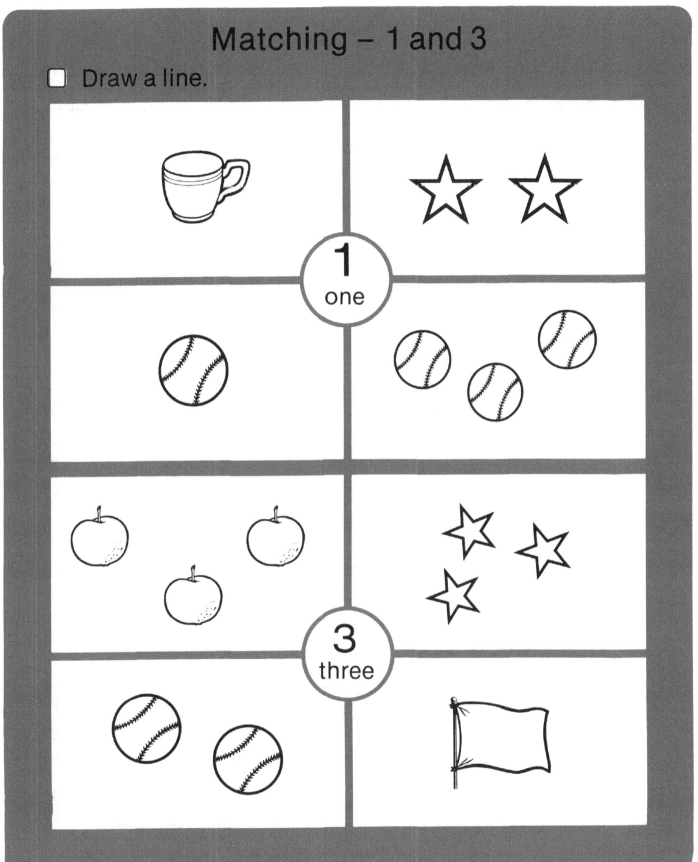

Matching – 2 and 3

☐ Draw a line.

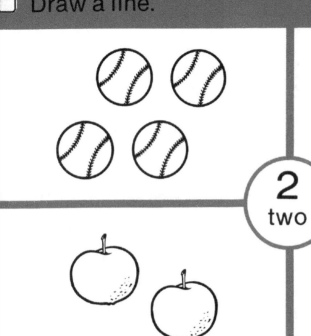

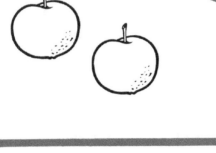

2 two

3 three

Matching – 1 to 3

☐ How many?

1   2   3

1   2   3

1   2   3

1   2   3

✓ Matching – 1 to 3

☐ How many?

1   2   3

1   2   3

1   2   3

1   2   3

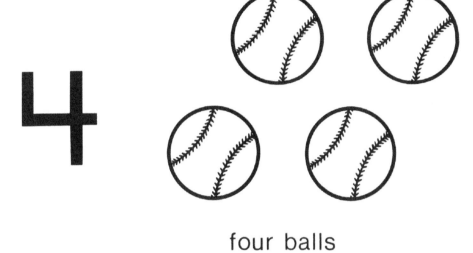

Color.

4 – four

☐ Color 4 balls.

4

four balls

☐ Color 4 cups.

4

four cups

15

Color.

☐ Color 4 shoes.

4

four shoes

☐ Color 4 apples.

 4

four apples

Color.

5 – five

☐ Color 5 balls.

5

five balls

☐ Color 5 cups.

5

five cups

Name _____ Date _____

Color.

☐ Color 5 shoes.

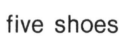

five shoes

5

☐ Color 5 apples.

5

five apples

18

Name _____   Date _____

Matching – 4 and 5

☐ Draw a line.

4
four

5
five

☐ Draw a line.

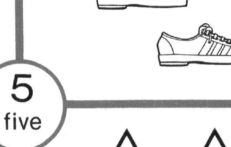

5 five

4 four

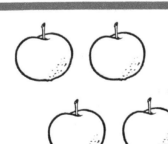

Matching – 1 to 5

☐ How many?

1　2　3

3　4　5

3　4　5

1　2　3

2　3　4

3　4　5

✓ Matching – 1 to 5

☐ How many?

1 2 3

2 3 4

3 4 5

1 2 3

2 3 4

3 4 5

Trace.

Writing 1

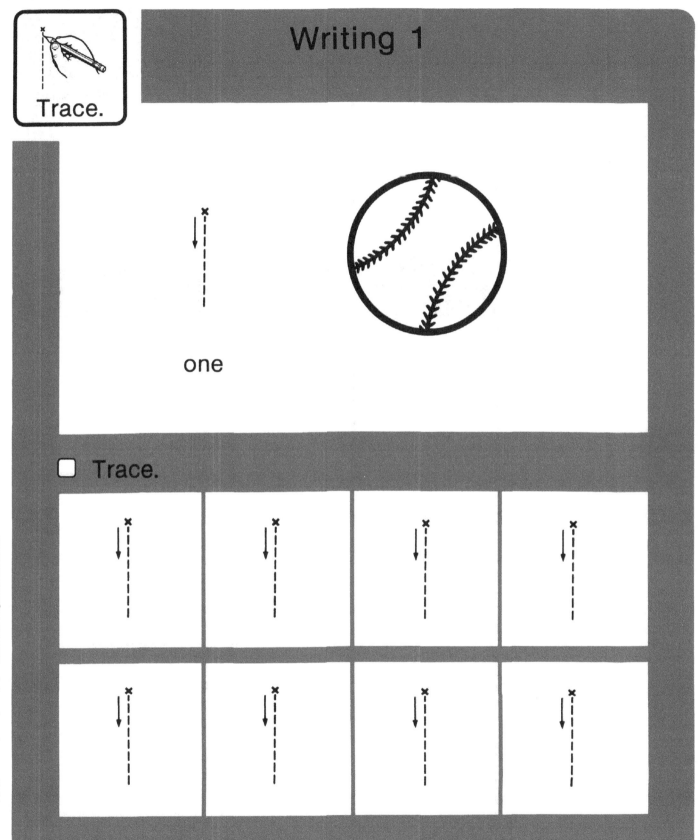

one

☐ Trace.

Name _____ Date _____

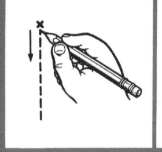 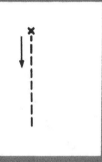 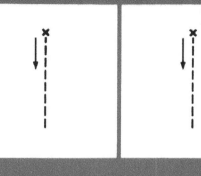

☐ Write.

Trace.

Writing 2

2

two

☐ Trace.

 2 2 2

 2 2 2

☐ Trace.

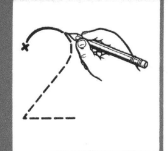

2 2 2

☐ Write.

✗ ✗ ✗

✗ ✗ ✗ ✗

✗ ✗ ✗ ✗

Writing 3

☐ Trace.

three

☐ Trace.

☐ Trace.

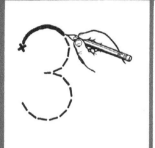   

☐ Write.

Writing 1 to 3

☐ How many?

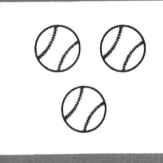			
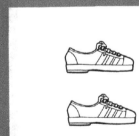			

29

Counting 1 to 3

☐ How many?

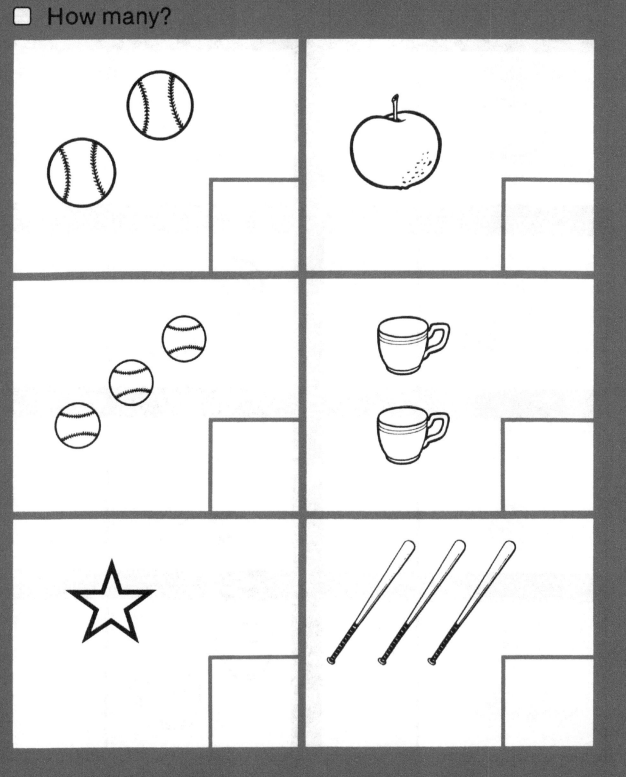

Writing 4

☐ Trace.

four

☐ Trace.

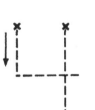 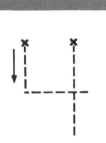 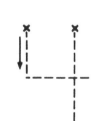

Name _____ Date _____

☐ Trace.

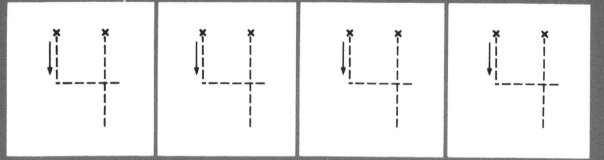

☐ Write.

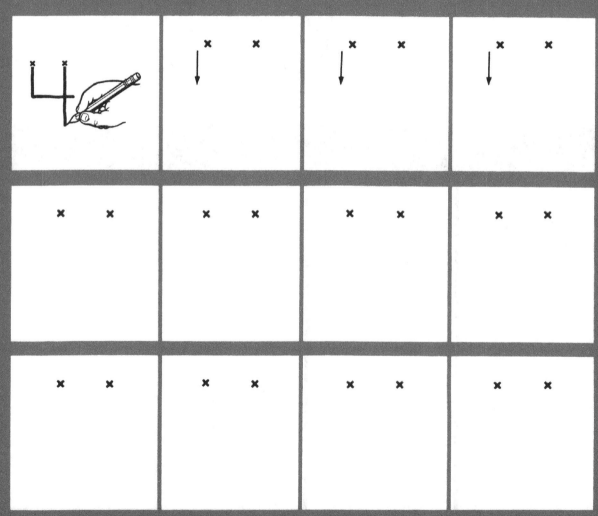

Writing 5

☐ Trace.

five

☐ Trace.

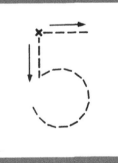

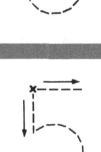
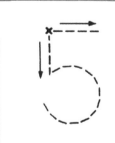

☐ Trace.

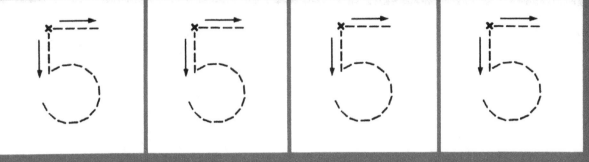

☐ Write.

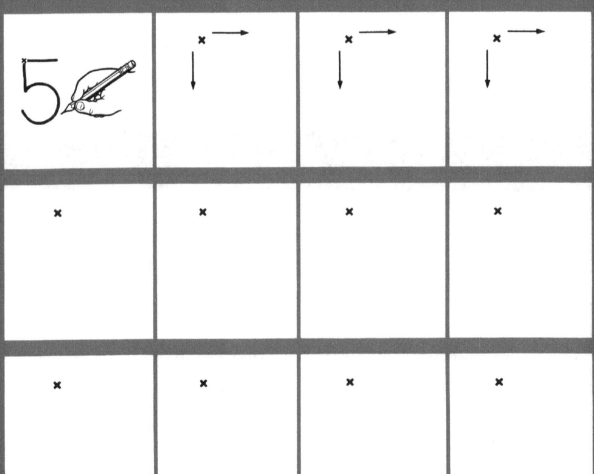

Writing 1 to 5

☐ Trace and write.

		✗	✗
		✗	✗
		✗	✗
	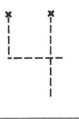	✗ ✗	✗ ✗
		✗	✗

✓ Writing 1 to 5

☐ Trace and write.

1	2	3	4	5
1	2	3	4	5
1	2	3	4	5
1	2		4	

Name _____ Date _____

Ⓡ Matching – 2 and 5

☐ Draw a line.

2
two

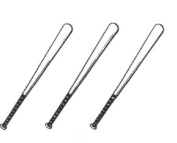

5
five

Ⓡ Matching – 3 and 4

☐ Draw a line.

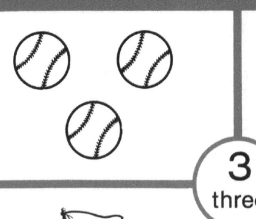

3
three

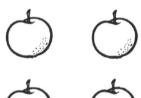

4
four

®Matching – 1 to 5

☐ How many?

1 2 3

1 2 3

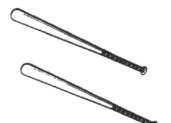

2 3 4

2 3 4

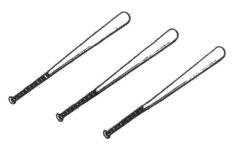

1 2 3

3 4 5

Ⓡ Counting 1 to 5

☐ How many?

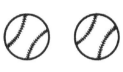

Name _____ Date _____

✓ Matching – 1 to 5

☐ How many?

1   2   3

2   3   4

3   4   5

1   2   3

2   3   4

3   4   5

✓Counting 1 to 5

☐ How many?

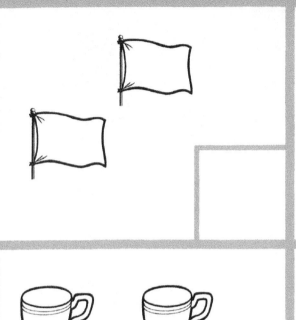

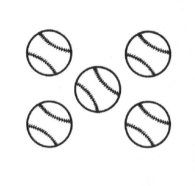

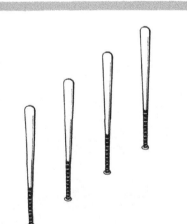

Ⓡ Matching – 1 to 5

☐ How many?

3   4   5

1   2   3

1   2   3

3   4   5

1   2   3

3   4   5

Ⓡ Counting 1 to 5

☐ How many?

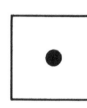

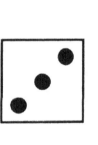

Name _____ Date _____

Counting 1 to 5

☐ Color.

1	
2	
3	
4	
5	

☐ Color.

3	
1	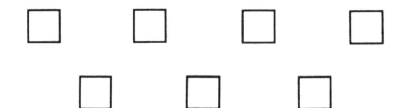
5	
2	
4	

Counting 1 to 5

☐ Draw balls.

1	◯
2	◯ ◯
3	
4	
5	

Counting 1 to 5

☐ Draw balls.

3	
5	
1	
4	
2	

Name _____ Date _____

✓ Counting 1 to 5

☐ **How many?**

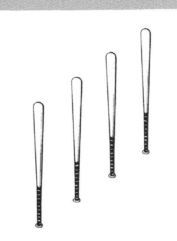

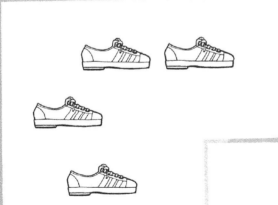

Name _____ Date _____

✓ Counting 1 to 5

☐ How many?

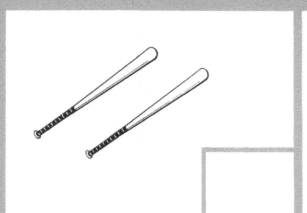

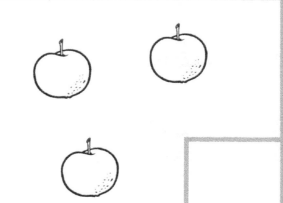

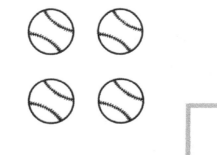

Name _____ Date _____

Writing 1 to 5

Trace and write.

●	1		
● ●	2		
● ● ●	3		
● ● ● ●	4		
● ● ● ● ●	5		

51

✓ Writing 1 to 5

☐ Trace and write.

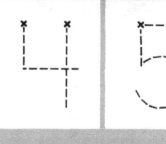

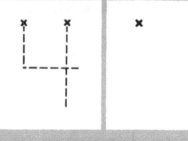

Color.

6 – six

☐ Color 6 balls.

six balls

☐ Color 6 boxes.

six boxes

☐ Draw 6 balls.

☐ Draw 6 boxes.

Name _____ Date _____

Color.

7 – seven

☐ Color 7 balls.

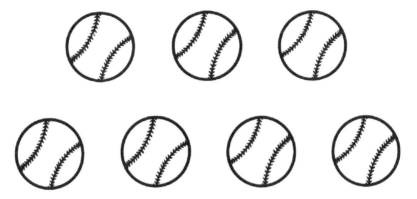

seven balls

☐ Color 7 boxes.

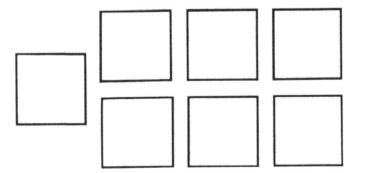

seven boxes

55

☐ Draw 7 balls.

☐ Draw 7 boxes.

Writing 6

☐ Trace.

six

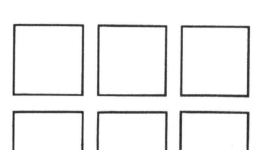

☐ Trace.

☐ Trace.

☐ Write.

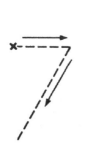

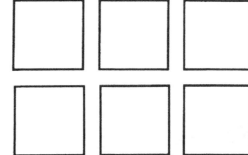

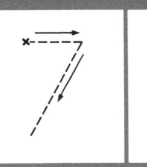

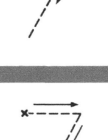

Name _____ Date _____

Writing 7

☐ Trace.

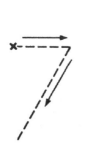 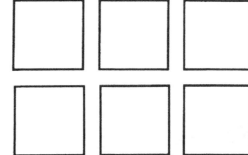

seven

☐ Trace.

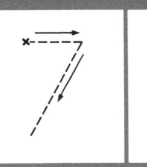 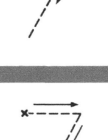

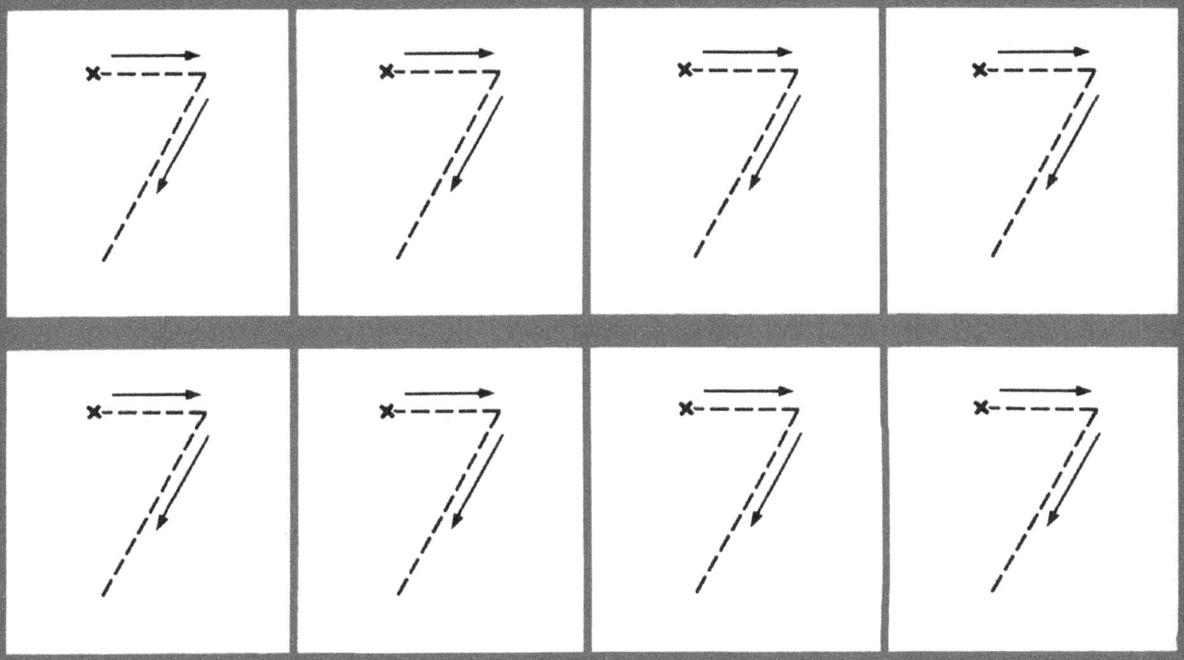

☐ Trace.

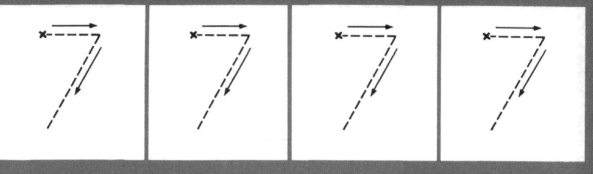

☐ Write.

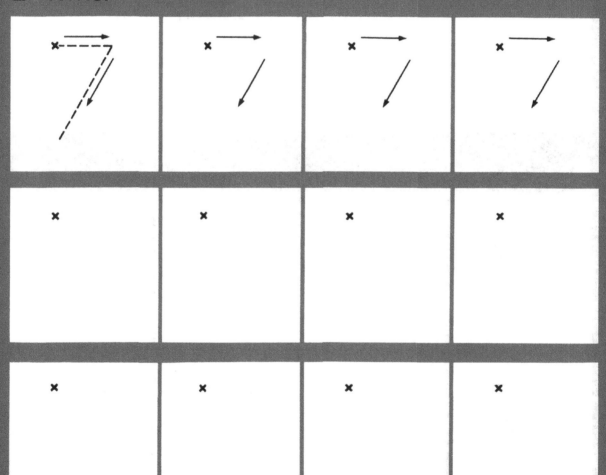

Matching – 5 to 7

☐ How many?

5   6   7

△ △
△ △ △
△ △

5   6   7

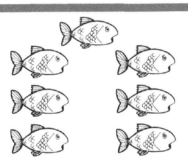

5   6   7

5   6   7

5   6   7

5   6   7

✓ Counting 5 to 7

☐ How many?

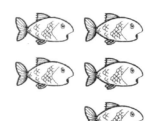

Writing 5 to 7

☐ How many?

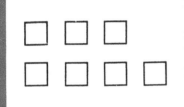   

  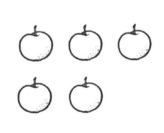

✓ Writing 3 to 7

☐ Trace and write.

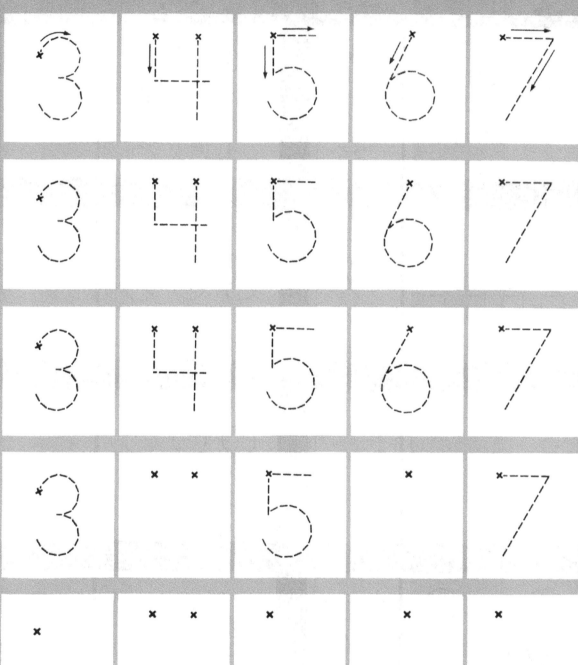

Name _____ Date _____

 Color.

8 – eight

☐ Color 8 balls.

eight balls

☐ Color 8 boxes.

eight boxes

Name _____ Date _____

☐ Draw 8 balls.

☐ Draw 8 boxes.

66

Name _____ Date _____

Color.

9 – nine

☐ Color 9 balls.

nine balls

☐ Color 9 boxes.

nine boxes

67

☐ Draw 9 balls.

☐ Draw 9 boxes.

Writing 8

☐ Trace.

eight

☐ Trace.

Copyright © 2002, 1989, 1980, by Phoenix Learning Resources, Inc.

69

☐ Trace.

☐ Write.

  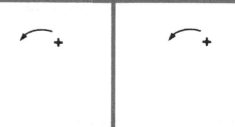

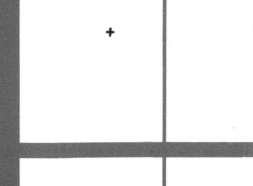

+ | + | + | +

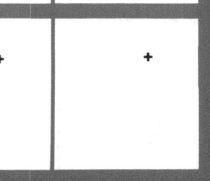

+ | + | + | +

Writing 9

☐ Trace.

nine

☐ Trace.

Name _____ Date _____

☐ Trace.

☐ Write.

| × | × | × | × |

| × | × | × | × |

Name _____ Date _____

Matching – 7 to 9

☐ **How many?**

7 8 9

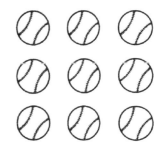

7 8 9

7 8 9

△ △ △
△ △
△ △

7 8 9

7 8 9

☐ ☐ ☐
☐ ☐ ☐
☐ ☐ ☐

7 8 9

73

Name _____ Date _____

✓ Counting 7 to 9

☐ **How many?**

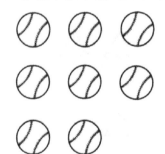

☐ ☐ ☐
☐ ☐ ☐
☐ ☐
☐

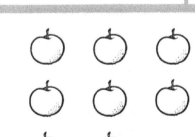

74

Writing 5 to 9

☐ Trace and write.

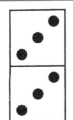			
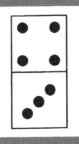			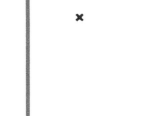
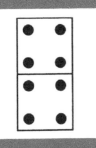		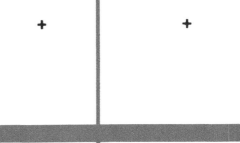	
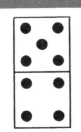		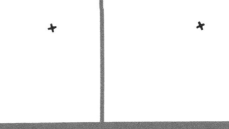	

Name _____ Date _____

✓ Writing 5 to 9

☐ Trace and write.

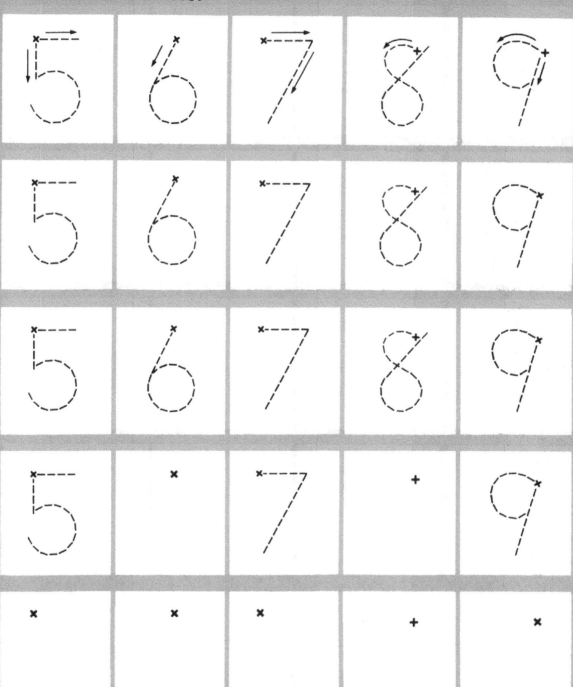

Color.

10 – ten

☐ Color 10 balls.

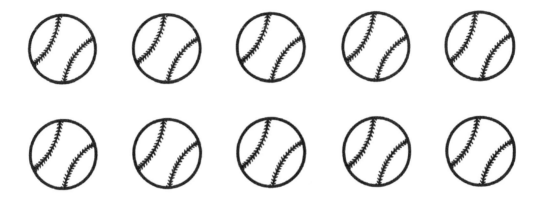

ten balls

☐ Color 10 boxes.

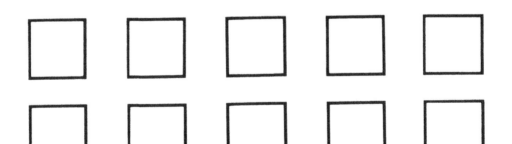

ten boxes

☐ Draw 10 balls.

☐ Draw 10 boxes.

Writing 10

☐ Trace.

10

ten

☐ Trace.

□ Trace.

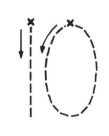 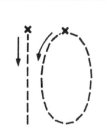 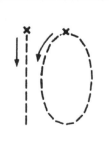 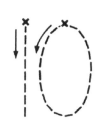

□ Write.

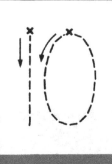

Matching – 8 to 10

☐ How many?

8 9 10

☐ ☐ ☐
☐ ☐ ☐ ☐
☐ ☐ ☐

8 9 10

8 9 10

△ △
△ △ △
△ △ △

8 9 10

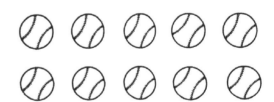

8 9 10

○ ○ ○ ○ ○
○ ○ ○ ○ ○

8 9 10

☐ How many?

Writing 7 to 10

☐ **How many?**

 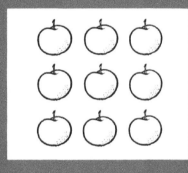 

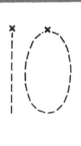 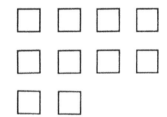 

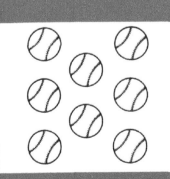  

✓ Writing 6 to 10

☐ Trace and write.

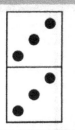		×	×
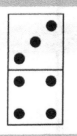		×	×
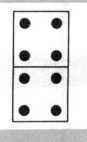		×	×
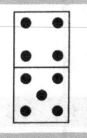		×	×
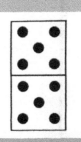	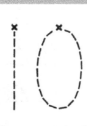	× ×	× ×

Name _____ Date _____

Matching – 6 to 10

☐ How many?

8 9 10

6 7 8

8 9 10

6 7 8

8 9 10

6 7 8

Counting 6 to 10

☐ How many?

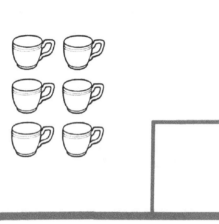

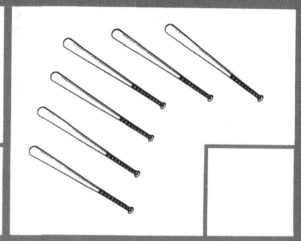

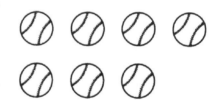

Name _____ Date _____

Counting 6 to 10

☐ Color.

6

7

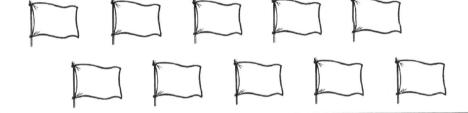

10

9

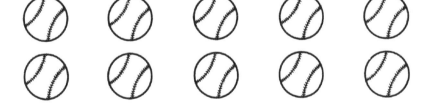

8

Counting 6 to 10

☐ Draw balls.

8	
7	
6	
9	
10	

✓ Counting 1 to 10

☐ Color.

2	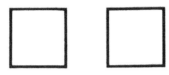
8	
5	
10	
7	

Name _____ Date _____

✓ Counting 1 to 10

☐ Color.

6	△ △ △ △ △ △ △ △
1	(5 flags)
3	(4 trees)
9	(10 diamonds)
4	(4 apples)

Matching – 1 to 5

☐ How many?

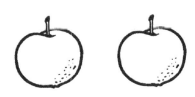

1 2 3

3 4 5

1 2 3

1 2 3

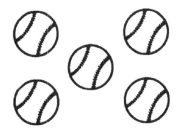

3 4 5

3 4 5

R Matching – 1 to 5

☐ How many?

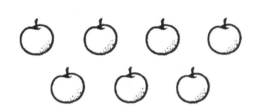

6 7 8

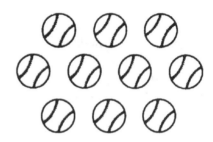

8 9 10

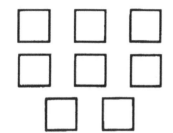

6 7 8

△ △ △
△ △ △

6 7 8

△ △ △ △
△ △ △ △

8 9 10

8 9 10

Counting 1 to 5

☐ **How many?**

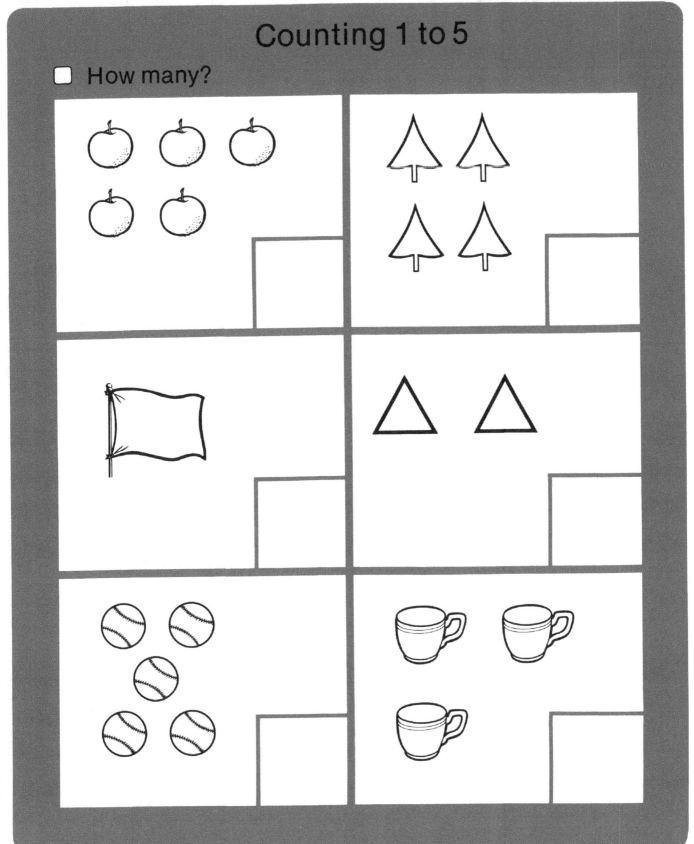

Counting 6 to 10

☐ How many?

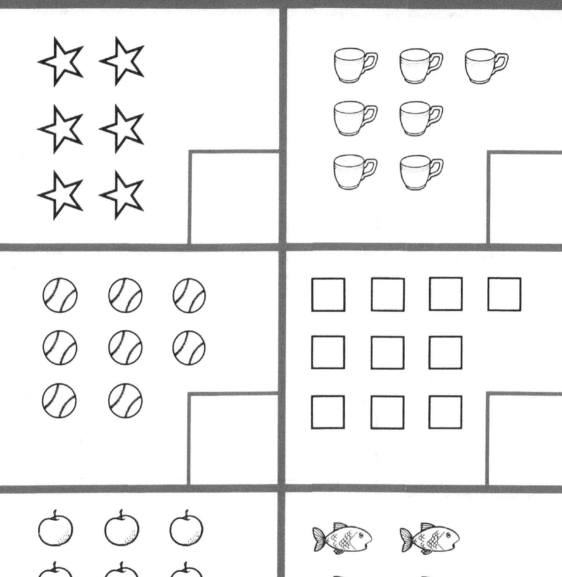

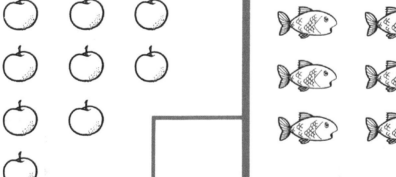

Writing 1 to 5

☐ Trace and write.

1	2	3	4	5
1		3		5
	2			

95

Writing 6 to 10

☐ Trace and write.

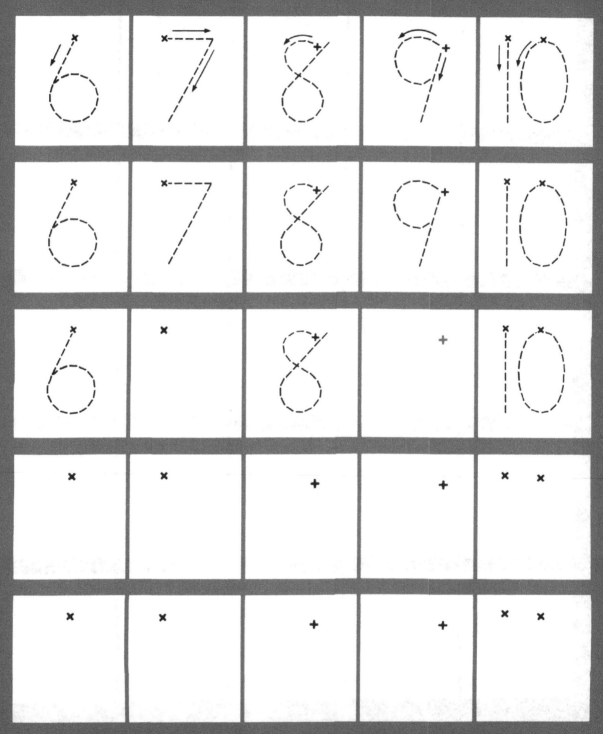

Name _____ Date _____

Ⓡ Writing 1 to 5

☐ Write.

◻	✖		
◻ ◻	✖		
◻ ◻ ◻	✖		
◻ ◻ ◻ ◻	✖ ✖		
◻ ◻ ◻ ◻ ◻	✖		

Ⓡ Writing 6 to 10

☐ Write.

⬛⬛ ⬛⬛ ⬛⬛	×		
⬛ ⬛ ⬛⬛⬛ ⬛ ⬛	×		
⬛ ⬛ ⬛⬛⬛ ⬛⬛⬛	×		
⬛⬛⬛ ⬛⬛⬛ ⬛⬛⬛	×		
⬛⬛⬛ ⬛⬛⬛⬛ ⬛⬛⬛	× ×		

Counting 1 to 5

☐ Color.

3	
1	
4	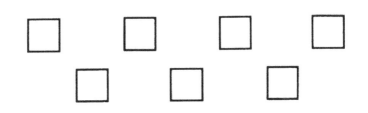
2	
5	

Counting 6 to 10

□ Color.

8	
6	
9	
7	
10	

✓ Counting 1 to 10

☐ Draw balls.

10	
4	
7	
2	
9	

✓ Counting 1 to 10

☐ Draw balls.

1	
8	
3	
6	
5	

Name _____ Date _____

Ⓡ Matching – 1 to 10

☐ How many?

2　3　4

5　6　7

3　4　5

1　2　3

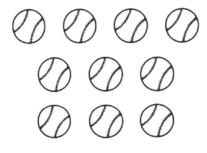

8　9　10

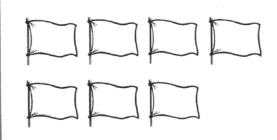

7　8　9

R Matching – 1 to 5

☐ How many?

2 3 4

3 4 5

5 6 7

6 7 8

8 9 10

4 5 6

R Counting 1 to 5

☐ Draw balls.

1	
5	
3	
2	
4	

Name _____ Date _____

Ⓡ Counting 6 to 10

☐ Draw balls.

9	
8	
7	
10	
6	

106

Name _____ Date _____

® Writing 1 to 10

■ Trace and write.

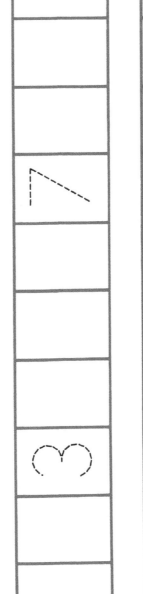
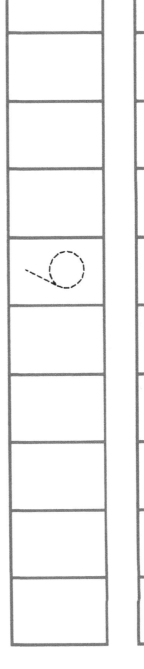
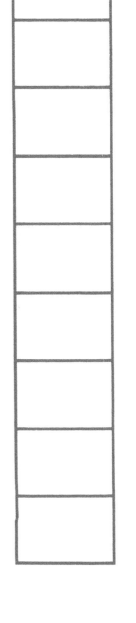

Name _____ Date _____

■ Writing 1 to 10

✓ Trace and write.

1	2	3	4	5	6	7	8	9	10
	2	5							
		6							

Name _____ Date _____

Ⓡ Counting 1 to 10

☐ **How many?**

109

✓ Counting 1 to 10

☐ How many?

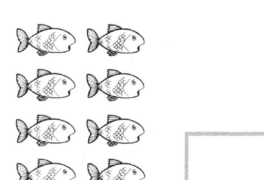

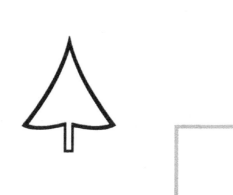

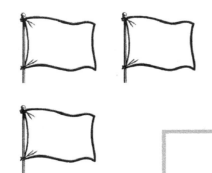